Beneath Earth's Surface

Greg Roza

Rosen
REAL
READERS

The Rosen Publishing Group, Inc.
New York

Published in 2001 by The Rosen Publishing Group, Inc.
29 East 21st Street, New York, NY 10010

Copyright © 2001 by The Rosen Publishing Group, Inc.

Book Design: Haley Wilson

ISBN: 0-8239-8155-X
6-pack ISBN: 0-8239-8557-1

Manufactured in the United States of America

Contents

Earth and the Solar System   5

The Layers of Earth   6

Under Our Feet   9

What's Inside the Crust?   15

Water and Air   19

Helping Earth   22

Glossary   23

Index   24

Neptune

Pluto

Uranus

Saturn

Jupiter

Mars

Earth

Venus

Mercury

4

Earth and the Solar System

We live on the **planet** Earth. It is part of a **solar system**. Our solar system is made up of the sun, Earth, eight other planets, moons, and other space objects. Everything in the solar system travels around the sun.

Earth is the third planet from the sun. It takes 365 days, or one year, for Earth to travel around the sun. Earth began forming about 4.5 billion years ago. After the sun was formed, leftover gas and dust began to circle it. Some of this gas and dust slowly came together to form Earth. Earth is special because it is the only planet we know of that has life. Plants and animals live almost everywhere on Earth's surface.

The names of the nine planets in our solar system are Mercury, Venus, Earth, Mars, Jupiter, Saturn, Uranus, Neptune, and Pluto.

The Layers of Earth

Earth has many **layers**. Some layers are solid, such as Earth's crust. Some layers are liquid, such as the oceans. Some layers are gas, such as the sky.

When Earth was forming, **gravity** pulled heavier things, like solids and liquids, toward the middle of Earth. This is why Earth has solid and liquid layers first and then a layer of gases.

Earth's first layer, starting at its center, is called the inner core. This is followed by the outer core. Around the core is the inner mantle. The outer mantle comes next. Around the mantle is an outer shell called the crust. The crust is what we live on. It is also where the oceans lie. The last layer of Earth is the air. The air is what we breathe every day.

We walk on top of many layers of Earth. The layer we live on is called the crust.

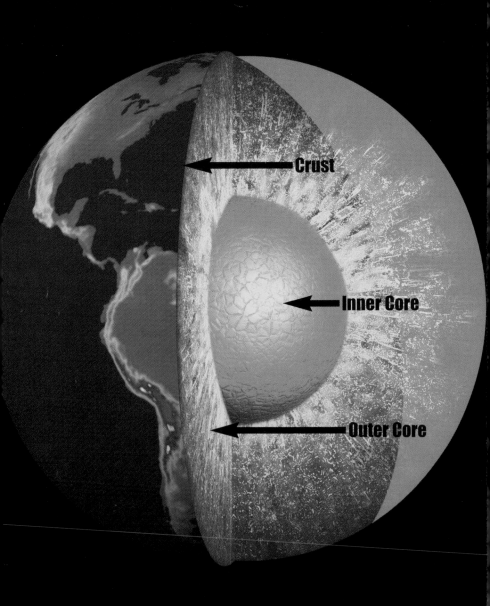

Crust

Inner Core

Outer Core

Under Our Feet

Earth's core, or center, begins about 1,800 miles below Earth's surface. If you could drive there, it would take about thirty-six hours. The very center of Earth is about 4,000 miles below the surface.

The core has two layers: the inner core and the outer core. The inner core is solid, and the outer core is hot liquid. The outer core is about twenty times larger than the inner core. Scientists believe the outer core may be about 1,400 miles thick. The distance from the outer core to the center of Earth may be about 800 miles. Both layers are made mostly of a metal called iron. The temperature of Earth's core ranges from about 8,000 to 11,000 degrees! We don't feel that heat because Earth's thick layers keep most of it from reaching the surface.

The solid inner core at the center of Earth is surrounded by a liquid outer core.

The mantle rests on top of the outer core. It is between six and forty miles below Earth's crust, and it is about 1,800 miles thick. The mantle is the thickest layer of Earth, making up about eighty percent of Earth. It is made up of iron and rock. The heat from the core makes the mantle very hot. It is about 6,700 degrees in the deepest part of the mantle, and about 1,800 degrees at the top.

The mantle has two areas: the lower mantle and the upper mantle. The lower mantle is solid rock. The upper mantle is also solid rock, but it has a thin outer layer that is part liquid. This liquid layer moves very slowly. This causes Earth's surface to change and move, but it moves much too slowly for us to feel it.

The upper mantle is very much like plastic. It is a solid that can flow or move when it gets very hot.

Upper Mantle

Lower Mantle

Crust

Lower Mantle

Upper Mantle

We live on Earth's crust, along with billions of plants and animals. Earth's crust is made of many different kinds of rocks. The upper layer of the crust can be anywhere from twenty to forty miles thick. Most of the heat from inside Earth does not reach the surface where we are. Some heat rises from the crust through **volcanoes**. When this happens, melted rock pushes through the crust and escapes as lava. This lava then cools and forms new crust.

Earth's surface is always moving. The surface is made of about twenty separate pieces of crust called plates. The plates rest on top of the moving, liquid layer of the upper mantle. The plates only move about two to four inches a year.

Smoke, steam, and lava come out of volcanoes when they erupt.

What's Inside the Crust?

We use many things found inside Earth's crust. Oil, coal, and natural gas come from the crust. These three things are thought to be what is left of the plants and animals that lived long ago. Some people believe that coal comes from the remains of plants that died millions of years ago. Oil and natural gas may have come from animals that once lived in the oceans.

Gems are valuable stones that are found in Earth's crust. Gems are dug up, cut, and cleaned until they shine. Diamonds are one kind of gem. Diamonds are created when the **pressure** inside Earth pushes on coal for thousands of years. Diamonds form in Earth's upper mantle. They are brought up near Earth's surface when volcanoes **erupt**.

Coal and diamonds are mined from Earth's crust. Coal can be used to heat our homes. People often wear diamonds in rings.

Many fossils are found in Earth's crust. A fossil is a rock formed from the bones of animals or parts of plants that were alive thousands or even millions of years ago. Fossils were formed from plants or animal bones that were quickly buried in the mud or sand at the bottom of rivers, lakes, swamps, and oceans. Over thousands of years, the pressure of the upper layers of the crust pressing down on the lower layers turned the plant and animal remains into rock.

Some fossils are leaves, shells, or skeletons that were left after a plant or animal died. Others are tracks or trails left by moving animals. Everything we know about dinosaurs we learned from the fossils they left behind.

Fossils can be found in Earth's crust in almost every state of the United States.

Water and Air

Did you know that three-quarters of Earth's crust is covered by ocean water? The planets nearest to us do not have any water on their surfaces, but there are two kinds of water on Earth.

The water in our oceans is called salt water because it has large amounts of salt in it. About ninety-seven percent of the water found on Earth is salt water. Oceans supply the water Earth needs for rainfall. If there were no oceans, life could not exist on our planet.

Rivers, streams, ice, and most lakes on Earth's surface are made up of freshwater. Three percent of Earth's water is freshwater. Freshwater is the water we use to drink and to bathe. Rain is the source of much of our freshwater.

Living things like fish, land animals, and plants need water to live.

When you look up at the sky, you are looking through a layer of gases that surrounds Earth. This is called the **atmosphere**. The gas that our bodies need to breathe is called **oxygen**. Oxygen makes up part of the atmosphere. We breathe this gas every day. The atmosphere also has a gas called **carbon dioxide**. Plants need carbon dioxide to make food for themselves.

The sun shines through the atmosphere, giving us heat and light. The atmosphere protects us from the sun's harmful rays. Clouds are also part of the atmosphere. Rain forms in clouds and falls to Earth, refilling the oceans, lakes, and rivers.

The atmosphere can be different colors depending on where the sun is, and how many clouds there are in the sky.

Helping Earth

We live on a very small part of Earth, but what we do can change everything on the planet. We need to remember how important our planet is to us. Some people have hurt Earth by cutting down too many trees in our forests. Others have **polluted** the air and water.

We all need to respect Earth and do what we can to keep our planet safe. You can help by learning to **recycle** things like cans and paper instead of throwing them away. The future of Earth is in our hands.

Glossary

atmosphere   The gases around Earth.

carbon dioxide   A gas that plants take in from the air and use to make food for themselves.

erupt   To burst or flow.

gravity   The force that causes objects to move toward the center of Earth.

layer   One thickness of something lying over or under another.

oxygen   A colorless gas that makes up part of the air we breathe.

planet   A large object that circles the sun.

pollute   To make air, water, or soil harmful to plants and animals.

pressure   A force that pushes on something.

recycle   To use again.

solar system   Our sun and the nine planets, moons, and other objects that circle it.

volcano   An opening in Earth's crust through which melted rock is sometimes forced.

Index

A
air, 6, 22
animal(s), 13, 15, 16

C
carbon dioxide, 20
coal, 15
core, 6, 9, 10
crust, 6, 10, 13, 15, 16, 19

F
fossil(s), 16

G
gas(es), 5, 6, 15, 20
gravity, 6

H
heat, 9, 10, 13, 20

L
liquid, 6, 9, 10, 13

M
mantle, 6, 10, 13, 15

O
ocean(s), 6, 15, 16, 19, 20

P
planet(s), 5, 19, 22
plant(s), 13, 15, 16, 20

R
rock(s), 10, 13, 16

S
sky, 6, 20
solar system, 5
solid(s), 6, 9, 10
sun, 5, 20

T
temperature, 9

V
volcanoes, 13, 15

W
water, 19, 22